Abdelhafid Mimouni

Mastering Dislocations in Materials

Abdelhafid Mimouni

Mastering Dislocations in Materials

ScienciaScripts

Imprint
Any brand names and product names mentioned in this book are subject to trademark, brand or patent protection and are trademarks or registered trademarks of their respective holders. The use of brand names, product names, common names, trade names, product descriptions etc. even without a particular marking in this work is in no way to be construed to mean that such names may be regarded as unrestricted in respect of trademark and brand protection legislation and could thus be used by anyone.

Cover image: www.ingimage.com

This book is a translation from the original published under ISBN 978-620-6-72306-6.

Publisher:
Sciencia Scripts
is a trademark of
Dodo Books Indian Ocean Ltd. and OmniScriptum S.R.L publishing group

120 High Road, East Finchley, London, N2 9ED, United Kingdom
Str. Armeneasca 28/1, office 1, Chisinau MD-2012, Republic of Moldova, Europe
Printed at: see last page
ISBN: 978-620-6-14592-9

Mastering Dislocations in Materials

Author: Dr. Abdelhafid Mimouni

An independent researcher in bioinorganic chemistry, Dr. Mimouni is an expert in macromolecular synthesis and characterization. He obtained his Ph.D. in Chemistry from the University of Paris XII in 1997, after a Diplôme des Études Approfondies en Systèmes Bioinorganiques at the University of Paris XI in 1993, where he also obtained his B.Sc. and M.Sc. in Chemistry.

Summary: This book explores in depth the role of dislocations in materials, covering their historical discovery, their impact on mechanical properties, and their modeling. After an introduction to the definition and structure of dislocations, the text presents the fundamental theory and mechanisms of dislocation displacement, illustrated by practical examples. Observation techniques such as transmission electron microscopy and X-ray diffraction are described in detail. The book also looks at practical applications of dislocations in superalloys and nanostructured materials, highlighting specific case studies. The management of dislocations to optimize material properties is discussed, offering an overview of current advances and future challenges in the field.

CONTENTS

Introduction

1. Presentation of the subject

Dislocation is a fundamental concept in materials science, playing a crucial role in understanding the mechanical properties of solid materials. Dislocations are linear defects in the crystal lattice that enable the plastic deformation of materials, an essential phenomenon in many industrial and technological applications.

The Importance of Dislocation in Materials Science

The discovery of dislocations dates back to the early years of the twentieth century, and marked a major advance in the understanding of material deformation mechanisms. Prior to this discovery, theories of plastic deformation in metals were based mainly on the atomic plane slip model. According to this theory, metals deformed by the sliding of atomic planes over each other, without taking into account internal imperfections such as dislocations.

The concept of dislocation was introduced independently by **G.I. Taylor** and **E.O. Hall** in the 1950s. **Frank** and **Bilby** also made a significant contribution to this theory by developing the dislocation model. These scientists showed that dislocations are essential for understanding how materials deform under stress. Taylor and Hall demonstrated that dislocations allow materials to deform plastically at lower stresses than those predicted by earlier theories.

Dislocation theory has revealed that materials do not simply behave as perfect entities, but contain internal defects that influence their mechanical properties. This has enabled us to better explain observed behaviors, such as the increased ductility of metals and the impact of heat and mechanical treatments on their properties. This discovery has led to significant improvements in material design and processing, influencing sectors such as aeronautics, automotive and advanced technologies.

Applications and Impact on Material Properties

Dislocations influence several key properties of materials:

- **Resistance to fracture**: Dislocations affect the way materials break under load. Managing them enables us to design materials that are more resistant to fracture.
- **Ductility**: The presence of dislocations facilitates plastic deformation, increasing the ability of materials to deform before breaking.
- **Fatigue**: Dislocations play a crucial role in the development of cracks during repeated loading cycles, thus influencing the lifespan of materials subjected to fatigue conditions.

Understanding and managing dislocations is essential for optimizing the properties of materials used in critical applications, from structural components to advanced electronic devices.

2. Book objectives

Defining what readers can expect to learn

This book aims to provide an in-depth understanding of dislocations, from their fundamental theory to their practical applications. Readers will learn:

- The nature and types of dislocations, including wedge and helical dislocations.
- Dislocation movement mechanisms and their impact on material properties such as strength and ductility.
- Modern techniques for observing and modeling dislocations, including electron microscopy and computer simulations.
- Strategies for controlling dislocations in materials to improve their performance in specific applications, such as metal alloys and composite materials.

Target audience

This book is aimed at a wide range of readers:

- **For students**: Those taking courses in materials science or engineering will find detailed explanations and concrete examples to complement their academic training.
- **Researchers**: Materials science and engineering researchers will benefit from in-depth discussions of dislocation mechanisms and case studies illustrating the latest advances in the field.
- **Engineers**: Professionals involved in materials design and processing will be able to apply the knowledge gained to solve practical problems related to materials properties.

Chapter 1: Introduction to Dislocation

1. Definition of Dislocation

What is a Dislocation?

A dislocation is a linear defect in a crystal lattice where the atomic planes are not continuous. It is a disturbance in the regular lattice structure that facilitates the movement of atoms and thus the plastic deformation of materials. Unlike a point defect or crack, a dislocation is an extended line across the crystal, allowing materials to deform under less stress.

Types of dislocation: Edge dislocation and Screw dislocation

1. **Edge dislocation**: This type of dislocation is formed by an extra plane of atoms inserted into the crystal lattice. The dislocation creates a deformation that propagates along a line perpendicular to the axis of the dislocation. Atoms are pushed and pulled from their normal positions, creating a zone of compression and a zone of expansion around the dislocation.

2. **Screw dislocation**: In this case, the crystal lattice is deformed helically around the dislocation line. The deformation is parallel to the dislocation axis, creating a spiral structure. This type of dislocation is formed when a uniform offset occurs along the axis, resulting in helical twisting of the crystal lattice.

2. Dislocation structure

Description of Network Defects Associated with Dislocations

Dislocations modify the regular structure of the crystal lattice:

- **Wedge dislocation**: introduces an additional plane of atoms, creating a zone of tension and a zone of compression in the lattice. This leads to a local change in atomic density, causing deformation around the dislocation.
- **Helical dislocation**: Creates a spiral deformation around the dislocation line, shifting atomic planes continuously along the axis. This results in a progressive shift of the crystal lattice in a helical direction.

Schematics and Illustrations of Different Types of Dislocations

The diagrams below show the dislocation structures:

- **Wedge dislocation**: A typical diagram shows an additional plane of atoms inserted into the lattice, with zones of compression and expansion.

- **Helical dislocation**: A diagram showing the helical torsion of the crystal lattice around the dislocation line, illustrating the progressive displacement of atomic planes.

3. The importance of dislocations in materials

Influence on Mechanical Properties: Strength, Ductility

Dislocations play a crucial role in the mechanical properties of materials:

- **Strength**: The presence of dislocations enables materials to deform plastically at stresses lower than those required to displace a perfect plane of atoms. Consequently, materials with a high density of dislocations can be strengthened by mechanisms such as strain-hardening.
- **Ductility**: Dislocations facilitate the sliding of atomic planes, increasing the ability of materials to deform before breaking. This improves the ductility of materials, which is essential for their use in applications requiring high deformation without fracture.

Chapter 2: Dislocation theory and mechanisms

1. Dislocation theory

Frank and Bilby Dislocation Model

The dislocation model developed by **Frank** and **Bilby** in the 1950s represented a major advance in our understanding of dislocations. According to this model, a dislocation is described as a line in a crystal along which local deformations occur. The model stands out for its ability to explain how dislocations move and interact in a crystal lattice.

Frank and Bilby proposed that dislocations are associated with perturbations in the atomic structure of the crystal. Their model shows that dislocations allow atomic planes to slip more easily, which explains the plastic deformation of materials. They also introduced concepts such as the wedge dislocation and the helical dislocation, providing a basis for modern dislocation theory.

Basic equations and key concepts

Key concepts in dislocation theory include:

- **Dislocation tension**: The force per unit length acting along the dislocation line. Dislocation tension is responsible for the movement of dislocations through the crystal lattice.

- **Burgers Vector**: A vector that describes the amplitude and direction of the local offset in the crystal lattice around the dislocation. It is crucial for characterizing the nature and behavior of dislocations.
- **Deformation field**: The way in which the presence of a dislocation deforms the crystal lattice around the dislocation line. This includes zones of compression and expansion, which influence the material's mechanical properties.

The basic equations associated with dislocations include:

- **Dislocation Tension Equation**: This equation describes the relationship between the force acting on a dislocation and the Burgers vector.
- **Deformation Field Equation**: Provides information on how the deformation of the crystal lattice is distributed around the dislocation.

2. Dislocation displacement mechanisms

Dislocation slip and overflow

Dislocations move through the crystal lattice by two main mechanisms:

- **Slip**: The most common mechanism by which a dislocation moves along a slip plane. This movement is facilitated by the presence of dislocations, which reduces the stress required to move atomic planes.
- **Overflow**: This mechanism occurs when a dislocation crosses one slip plane and moves onto another, often perpendicular, plane. Overflow is important for the formation of complex structures and for the management of deformations in materials.

Role of Slip Plans and Slip Directions

Slip planes are specific planes in a crystal where dislocations move most easily. **Slip directions** are the directions in which dislocation movement is favored. These planes and directions are determined by the material's crystalline structure and strongly influence the way materials deform under stress.

- **Slip planes**: For example, in face-centered cubic crystals (FCC), slip planes are generally {111} planes. In body-centered cubic crystals (CBC), {110} planes are often the most favorable.
- **Slip directions**: Slip directions in CFC materials are often <111> directions, while in CBC materials, <111> directions are also favored.

3. Concrete examples

Case Study 1: Plastic deformation in metals

Example: How Dislocations Contribute to Plastic Deformation in Steel in a Tensile Test

In tensile testing of steel, dislocations play a crucial role in the way the material deforms. When steel is subjected to stress, dislocations move along slip planes, allowing the material to deform plastically before breaking. The density and distribution of dislocations directly influence the strength and ductility of steel. Heat treatments such as quenching can modify the density of dislocations, thus affecting the mechanical properties of steel.

Case Study 2: Dislocations in Semiconductors

Example: Impact of Dislocations on the Performance of Silicon Devices

In semiconductors such as silicon, dislocations can have a significant impact on the performance of electronic devices. Dislocations in silicon can cause defects in the crystal lattice that affect the behavior of charge carriers, leading to variations in the electrical characteristics of devices such as transistors. Managing dislocations through controlled crystal growth and heat treatment techniques is essential for manufacturing high-performance devices.

Chapter 3: Observation and characterization techniques

1. Transmission electron microscopy (TEM)

Methodology for observing dislocations

Transmission Electron Microscopy (TEM) is a key technique for observing dislocations at the atomic level. It uses electron beams transmitted through a very thin sample. The electrons interact with the atoms in the crystal lattice, creating contrasts that reveal the presence of dislocations. The dislocations appear as lines or contrast zones in the TEM images, enabling direct visualization of crystal lattice deformations.

Examples of MET images

TEM images show dislocations as distinct lines across the crystal lattice. For example, images can show wedge-shaped and helical dislocations, with sharp contrasts around the dislocation lines. These images enable the structure and density of dislocations to be visualized, providing crucial information on their distribution and impact on the material.

2. X-ray diffraction

How Diffraction Can Reveal the Presence and Movement of Dislocations

X-ray diffraction is an effective technique for studying the presence and movement of dislocations by analyzing how X-rays are diffracted by the crystal lattice. When dislocations are present, they create deviations in the diffraction planes, causing broadened diffraction peaks or changes in peak intensity. These variations provide clues to the density and movement of dislocations in the crystal.

Practical example: Analyzing the structure of a copper crystal

In the analysis of a copper crystal, X-ray diffraction can reveal information about the crystal structure in the presence of dislocations. Diffraction measurements show broadening of diffraction peaks, indicating the presence of dislocations. Using advanced diffraction techniques such as high-resolution X-ray diffraction, it is possible to quantify the density of dislocations and study their effect on copper properties.

3. Atomic Force Spectroscopy (AFM)

Using AFM to Study Surface Dislocations

Atomic Force Spectroscopy (AFM) is used to examine surface dislocations with nanometric resolution. AFM uses a probe that scans the surface of the material, detecting variations in the topography caused by dislocations. This technique provides three-dimensional images of the material surface, enabling dislocations and other surface defects to be identified and characterized.

Typical applications and results

AFM can be used to observe dislocations in samples of metals, semiconductors and other materials. For example, in silicon samples, AFM can reveal dislocations as irregularities in the surface, with defect heights corresponding to the dislocations. Typical results include detailed images showing the characteristics of dislocations, their density, and their distribution on the sample surface.

This chapter presents the main techniques used to observe and characterize dislocations, including TEM, X-ray diffraction and AFM, with concrete examples showing their application in various materials.

Chapter 4: Modeling and Simulation of Dislocations

1. Modeling Methods

Atomistic Modeling and Continuum Models

There are two main approaches to modeling dislocations: atomistic and continuous.

- **Atomistic modeling**: This method simulates dislocations by taking into account the interactions between individual atoms. Atomistic models use molecular dynamics simulations to study atomic motion and dislocation formation. These models provide a detailed understanding of the mechanisms at the atomic scale, and enable us to predict the properties of materials as a function of their atomic defects.
- **Continuous models**: In contrast, continuous models use approximations that treat dislocations as macroscopic entities in a continuum of matter. These models, such as dislocation elasticity theory, simplify atomistic complexity by focusing on the global effects of dislocations on mechanical properties. Continuum models are often used to predict material behavior on a larger scale, and are well suited to the analysis of large-scale deformations.

Software and Tools Used in Dislocation Simulation

Several software packages and tools are used to simulate dislocations:

- **LAMMPS (Large-scale Atomic/Molecular Massively Parallel Simulator)**: Molecular dynamics software for simulating complex atomistic systems, including dislocations.
- **ATAT (Automated Tool for Analysis of Transport)**: Used to model the structure and interactions of dislocations in materials.
- **ABINIT**: A simulation program for electronic structure calculations, including simulations of dislocations in materials.

These tools make it possible to model and simulate the behavior of dislocations in various materials, facilitating research into their effects on mechanical and structural properties.

2. Simulation applications

Prediction of Mechanical Properties of Materials

Dislocation simulations help predict the mechanical properties of materials by analyzing how dislocations influence their strength, ductility and other properties. For example, simulations can model how the introduction of dislocations affects the tensile strength of a metal alloy. By adjusting the parameters of the simulations, researchers can predict how different thermal or mechanical treatments modify material properties.

Practical example: Simulating the impact of dislocations on alloy hardness

For a specific alloy, simulations can show how dislocations affect hardness. For example, a simulation might show that increasing the density of dislocations increases the alloy's strength by hindering the movement of other dislocations, a phenomenon observed in dispersion-strengthened alloys. Simulations can also be used to optimize compositions and heat treatments to predictively improve alloy hardness.

3. Case studies

Case Study 1: Modeling Dislocations in an Aluminum Alloy

Example: How Predictive Simulations Help Design Stronger Alloys

In this study, simulations are used to model dislocations in an aluminum alloy. Using atomistic simulations and continuum models, researchers can observe how dislocations interact with other defects in the alloy. The results show how dislocation density and distribution affect the alloy's tensile strength and ductility. This information makes it possible to design aluminum alloys with improved mechanical properties, suitable for specific applications such as aerospace components.

Case Study 2: Effect of Dislocations on Material Fatigue

Example: Simulation of Fatigue Cycles and their Effects on Dislocations in a Construction Material

Background and objective

Construction materials are often subjected to cyclic loads, which can lead to fatigue phenomena. Dislocations have a significant influence on the fatigue behavior of materials, affecting crack formation and propagation. Simulations can be used to study these effects in detail and improve the durability of materials.

Simulation Methodology

1. **Fatigue Cycle Modeling**
    - **Model creation**: We start by creating a numerical model of the material, introducing dislocations. Either atomistic or continuous models can be used. For atomistic simulations, molecular dynamics is often employed, while continuous models can use dislocation elasticity theory.

- **Cyclic loading**: Loading cycles are simulated by imposing repeated stresses on the model. The shape of the load can be sinusoidal, square or triangular. The applied stress is generally described by:

$$\sigma(t) = \sigma_{max} \sin(2\pi ft)$$

Where $\sigma(t)$ is the stress at an instant t, σ_{max} is the maximum stress, f is the cycle frequency, and t is time.

2. **Observation of Dislocation Effects**

- **Dislocation evolution**: Simulations follow the evolution of dislocations under the effect of loading cycles. The evolution of dislocations is often modeled by equations that describe their motion and interaction with other dislocations. For example, the dislocation density equation ρ under stress can be given by:

$$\rho(t) = \rho_0 + \Delta\rho \sin(2\pi ft)$$

where ρ_0 is the initial dislocation density, and $\Delta \boldsymbol{\rho}$ is the variation in dislocation density due to cyclic loading.

- **Crack formation**: Simulations show how dislocations facilitate crack formation. Cracks propagate faster around dislocations, which can be modeled by the crack propagation equation:

$$da/dN = 1/(K_{max} - K)_{min}$$

Where a is the crack length, N is the number of cycles, and K_{max} and K_{min} are the maximum and minimum values of the stress field intensity.

Results and analysis

1. **Contribution of Dislocations to Fatigue**

   - **Crack propagation**: The results show that dislocations promote crack formation by increasing the concentration of local stresses. The speed of crack propagation is modeled by the Paris-Erdogan law:

$$da /dN = C(\Delta K)^m$$

where da/dN is the crack propagation velocity per cycle, ΔK is the amplitude of the stress factor, and C and m are material constants.

History of the Paris-Erdogan Law

The Paris-Erdogan law is a key relationship for predicting crack propagation speed under cyclic loading. It was developed independently by two researchers: Paul C. Paris and Fatih Erdogan.

1. **Paul C. Paris**:

   - **Biography and Contributions**: Paul C. Paris, aerospace engineer and researcher, was one of the pioneers in the study of crack propagation. He began his work on material fatigue in the 1960s, in response to challenges encountered in aeronautical structures.

   - **Key works**: Paris formulated an empirical law for crack propagation in response to experimental data. His research led to the development of the model that bears his name, based on the relationship between crack propagation speed and stress field intensity.

2. **Fatih Erdogan**:

   - **Biography and Contributions**: Fatih Erdogan, an engineer and researcher in mechanics of materials, collaborated with Paris to develop this law. Erdogan contributed to the theoretical analysis and interpretation of the mechanisms underlying crack propagation.

- **Key works**: Erdogan has made important contributions to the understanding of crack propagation mechanisms and the application of the Paris-Erdogan law in various material contexts and loading conditions.
- **Dislocation density**: A higher dislocation density leads to accelerated crack propagation. Simulations show that this density increases the material's failure rate.

2. **Implications for Materials Durability**
   - **Designing Resistant Materials**: By optimizing heat treatments and alloy compositions, we can control the density and distribution of dislocations, thereby improving fatigue resistance. For example, alloys with low dislocation density can have a longer service life under cyclic loading.
   - **Prediction methods**: Simulations can be used to predict the service life of materials as a function of loading conditions and the presence of dislocations. These predictions are essential for the design of durable structures and the planning of maintenance interventions.

Conclusion

Fatigue cycle simulations enable us to study in detail the effect of dislocations on the behavior of materials subjected to repeated loading. They offer valuable insights for designing more fatigue-resistant materials and improving methods for predicting their service life, thus contributing to the creation of safer, longer-lasting structures.

This chapter explores methods for modeling dislocations, applications of simulations to predict material properties, and presents case studies illustrating how these simulations contribute to materials design and analysis.

Chapter 5: Practical and advanced applications

1. Effect of Dislocations on Material Properties

Influence on Resistance to Rupture, Fatigue, and Plasticity

Dislocations play a crucial role in determining the mechanical properties of materials. Their presence and movement directly influence the fracture toughness, fatigue and plasticity of materials.

- **Fracture strength**: Dislocations affect the way a material fractures. The presence of dislocations can allow plastic deformation, which can actually improve fracture toughness by dissipating concentrated stresses and allowing more uniform deformation before fracture.
- **Fatigue**: Dislocations are also involved in the fatigue process. Under cyclic loading, dislocations can interact and accumulate, promoting crack formation and influencing material life under repeated loading.
- **Plasticity**: The plasticity of a material is enhanced by the presence of dislocations. Their movement allows the material to deform without breaking, thus increasing its ductility.

Practical examples from the aerospace and automotive industries

- **Aerospace industry**: In gas turbines, superalloys have to withstand extreme temperature and stress conditions. The management of dislocations in these alloys optimizes their performance, increasing their resistance to fatigue and fracture.

- **Automotive industry**: Materials used in high-performance components such as pistons and drive shafts also benefit from dislocation management. This improves their wear resistance and durability under severe operating conditions.

2. Dislocation Control and Optimization

Thermal and Mechanical Treatment Techniques for Dislocation Control

Dislocation control is essential for optimizing material properties. Heat and mechanical treatment techniques can be used to adjust the density and behavior of dislocations to achieve the desired properties.

- **Heat treatment**: Treatments such as quenching and tempering modify the dislocation structure and mechanical properties of materials. For example, high-temperature tempering after quenching reduces dislocation density and improves ductility while maintaining good strength.
- **Mechanical processing**: techniques such as rolling and extrusion modify the distribution and movement of dislocations. These treatments can be used to refine the material's microstructure, thereby increasing its strength and plasticity.

Examples of Optimization in Critical Component Manufacturing

- **Critical components**: In the manufacture of critical components, such as engine parts and high-performance structures, precise dislocation management is used to improve reliability and performance. For example, heat treatment of high-strength steels provides optimum characteristics for demanding applications.

3. Advanced Case Studies

Case Study 1: Dislocations in Superalloys

Background and importance of superalloys

Superalloys are critical materials used in extreme environments, such as aircraft engine gas turbines and power plant reactors. These alloys must maintain their mechanical properties at high temperatures, often in excess of 1000°C, while resisting corrosion and wear. Dislocations play a crucial role in the performance of superalloys, and their management is essential to optimize their resistance to fatigue and fracture.

Dislocation Management in Superalloys

Dislocation management in superalloys is achieved by a variety of means, including adjustment of chemical composition, heat treatment, and control of manufacturing conditions. Here's how these factors influence superalloy performance:

1. **Chemical composition**

   - **Adding alloying elements**: Elements such as molybdenum, tungsten and cobalt are added to superalloys to form precipitates that prevent the movement of dislocations. These precipitates, such as carbides and gamma' phases, act as barriers to dislocations, thus strengthening the alloy.

     Case in point: Inconel 718 superalloy, used in gas turbines, contains niobium and titanium which form gamma' precipitates. These precipitates prevent dislocation slippage, thus improving fracture toughness and fatigue strength at high temperatures.

2. **Heat treatment**

   - **Quenching and tempering**: Heat treatments are used to modify the size, shape and distribution of precipitates and dislocations. For example, tempering after quenching can adjust the density of dislocations and optimize the material's mechanical properties.

     Case in point: In the CMSX-4 superalloy, high-temperature heat treatment followed by rapid cooling can increase the density of gamma' precipitates while reducing the density of dislocations, thereby improving the material's fatigue strength and dimensional stability.

3. **Manufacturing conditions**

   - **Manufacturing processes**: Manufacturing processes such as melting, forging and extrusion influence the microstructure of superalloys, including the distribution of dislocations. Precise control of manufacturing conditions helps reduce internal defects and optimize material properties.

     Case in point: Rene 41 superalloy, used in gas turbines, is manufactured using a vacuum melting process to minimize impurities and control grain size. This controls dislocation density and maximizes the material's mechanical performance.

Concrete examples of applications and dislocation management

1. **Inconel 718 superalloy**

   - **Description**: Inconel 718 is a nickel-chromium superalloy used in gas turbines and aircraft engine components. It contains elements such as niobium and titanium, which form gamma' precipitates in the nickel matrix.

   - **Dislocation management**: The presence of gamma' precipitates prevents the movement of dislocations, thereby increasing the fracture toughness and durability of the superalloy at high temperatures. Heat treatment is used to adjust the size and distribution of precipitates, optimizing mechanical properties.

2. **CMSX-4 superalloy**
    - **Description**: CMSX-4 is a single-crystal superalloy used in gas turbines for high-temperature applications. It contains chromium and molybdenum, which form gamma' precipitates dispersed in the nickel matrix.
    - **Dislocation management**: Carefully controlled heat treatment adjusts the density of dislocations and precipitates, improving fatigue strength and overall superalloy performance. Mechanical properties are optimized by high-temperature tempering, which controls precipitate size and dislocation density.
3. **Superalloy Rene 41**
    - **Description**: Rene 41 is a nickel-based superalloy used in gas turbine components. It is known for its good corrosion resistance and ability to maintain its mechanical properties at high temperatures.

- **Dislocation management**: Rene 41 is manufactured using vacuum melting and heat treatment to control the density of dislocations and precipitates. This optimized dislocation management contributes to improved mechanical performance and extended component life.

Conclusion

Dislocation management in superalloys is essential to optimize their mechanical properties and performance under extreme conditions. By adjusting chemical composition, applying heat treatments, and controlling manufacturing conditions, it is possible to manipulate the density and distribution of dislocations to improve fatigue and fracture resistance. Concrete examples of superalloys such as Inconel 718, CMSX-4 and Rene 41 illustrate how dislocation management can be used to develop high-performance materials for critical applications.

Case Study 2: Dislocations in Nanostructured Materials

Background and importance of nanostructured materials

Nanostructured materials, with dimensions in the nanometer range, have unique physical and mechanical properties that differ from those of materials on the macroscopic scale. At this scale, surface effects, changes in crystalline structure and dislocation interactions become predominant. Metal nanowires, nanotubes and nanoparticles illustrate how mechanical properties can be modified by dislocation management, influencing characteristics such as hardness, fracture toughness and ductility.

The Role of Dislocations in Nanostructured Materials

1. **Moving Dislocations**
   - **Increased mobility**: In metal nanowires, dislocations can move more easily due to the reduced sizes and increased stresses at the nanoscale. This phenomenon can improve the material's ductility and toughness, in contrast to larger-scale materials where dislocations can lead to faster fracture.

     Case in point: Copper nanowires show better ductility than conventional copper wires due to the facilitated movement of dislocations at the nanoscale. Studies show that the high density of dislocations in nanowires may contribute to a greater capacity for deformation without fracture.
2. **Surface Effects and Dislocation Control**
   - **Influence of surfaces and interfaces**: At the nanoscale, surfaces and interfaces have a significant impact on dislocation

dynamics. Nanostructured materials may feature interfaces between grains or between different phases that affect the way dislocations propagate and interact.

Case in point: In carbon nanotubes, dislocations and lattice defects are often observed along tube interfaces. These interactions can influence the mechanical and electrical properties of nanotubes, affecting their use in electronic and composite applications.

3. **Nanoscale Dislocation Modeling**

   - **Simulation approaches**: Atomistic simulations, such as the molecular dynamics method, are used to study the behavior of dislocations in nanostructured materials. These simulations provide information on how dislocations interact with surfaces and interfaces at the nanoscale.

     Case in point: Molecular dynamics simulations of tungsten nanowires show how dislocations interact with interfaces between grains and surfaces. The results reveal that dislocations can cluster together and cause localized deformations, which can be exploited to design nanostructured materials with specific mechanical properties.

Applications and Optimization of Nanostructured Materials

1. **Metal Nanowires**

    - **Mechanical properties**: Metal nanowires, such as copper or gold nanowires, exhibit increased strength and improved ductility compared to conventional wires due to the interactions of dislocations at the nanoscale.

      Case in point: gold nanowires used in sensors and electronic devices benefit from their ability to deform without breaking, which is partly due to the management of dislocations at the nanoscale. This makes it possible to design more robust and reliable devices.

2. **Carbon nanotubes**

    - **Electronic applications**: Carbon nanotubes, with their hexagonal structures and exceptional properties, use dislocations to enhance the performance of electronic devices. Dislocations in nanotubes can influence electrical conductivity and mechanical strength.

Case in point: Devices based on carbon nanotubes, such as field effect transistors (FETs), benefit from dislocation management to optimize their electronic performance. Understanding dislocation interactions helps to improve device stability and efficiency.

3. **Nanostructured Materials for Biomedical Applications**

    - **Improved properties**: Nanostructured materials are also used in biomedical applications, where dislocations can influence the biocompatibility and durability of implants and devices.

      Case in point: nanostructured titanium implants use dislocation management to improve fatigue resistance and durability in biological environments, offering more reliable implants for medical applications.

Chapter 6: The Legacy of Sir Alan Cottrell: Dislocations and Practical Applications

1. Bibliography of Sir Alan Cottrell

Sir Alan Cottrell (1923-2012) is widely recognized for his groundbreaking contributions to materials science, particularly in the field of dislocations. His work laid essential foundations for understanding the behavior of materials under stress. Here is a selection of his most influential publications:

- **Cottrell, A. H. (1953).** *Dislocations and Plastic Flow in Crystals.* Oxford University Press. This pioneering text presents the fundamental principles of the dislocation model and its role in the plastic flow of crystals.

- **Cottrell, A. H. (1964).** *The Strength of Solids*. Butterworths. In this book, Cottrell develops his theories on the strength of solids, with particular attention to the interactions between dislocations and other defects.
- **Cottrell, A. H. (1970).** *The Mechanical Properties of Materials*. Macmillan. This work explores in depth the mechanical properties of materials, incorporating discoveries about dislocations and their influence on the strength and ductility of materials.
- **Cottrell, A. H., & Oakeshott, L. H. H. B. (1980).** *The Theory of Crystal Dislocations*. Academic Press. This publication offers a comprehensive review of the theories developed on dislocations and their impact on material properties.

2. Discovery and Theories of Sir Alan Cottrell

Historical background:

In the 1950s, understanding of plastic deformation phenomena in metals was still limited. The theory of dislocations, introduced by Cottrell, provided an innovative explanation of the mechanical behavior of materials. Prior to his contributions, plastic deformation was mainly explained by sliding models of atomic planes, without taking complex internal defects into account.

Discovery:

Sir Alan Cottrell introduced key concepts about dislocations, the linear defects in the crystalline structure of materials. He formulated theories explaining how dislocations influence the way materials deform under stress, particularly at high temperatures.

Dislocation model:

- **Dislocation model:** Cottrell's dislocation model describes how dislocations move through crystal lattices and interact with neighboring atoms. He proposed that dislocations can multiply and form complex networks, influencing the mechanical properties of materials. This model is crucial for understanding plastic deformation mechanisms.
- **Effect of impurities:** Cottrell also developed the idea that dislocations can bind with impurities in metals, increasing the strength of the material by preventing dislocation movement. This interaction between dislocations and impurities is essential to explain the mechanical behaviors observed in metal alloys.

3. Key contributions and impacts

Theoretical model:

Cottrell's theories led to a better understanding of plastic deformation mechanisms. His theoretical model highlighted how dislocations move, lock and interact with other defects and impurities, offering a quantitative explanation of the strength and ductility of materials.

Practical applications:

- **Heat treatment:** Cottrell's discoveries have led to significant improvements in heat treatment processes for metal alloys. By understanding how dislocations interact with impurity atoms, engineers have been able to optimize the mechanical properties of alloys for critical applications.
- **Aerospace industry:** His theories have had a major impact on the aerospace industry. Metal alloys used in aircraft engines and other high-performance applications benefit from the principles developed by Cottrell, improving their resistance to fatigue and fracture.

Technological advances:

Cottrell's concepts have been incorporated into modern materials design, influencing fields such as alloy metallurgy, composite materials science and advanced technologies. His theories continue to be used to develop stronger, higher-performance materials that meet the demands of cutting-edge technologies.

4. Case studies and applications

Superalloys:

Superalloys, such as those used in gas turbines, illustrate the application of Cottrell's principles. By adjusting chemical composition and heat treatments, engineers can control the size and distribution of dislocations to optimize material performance at high temperatures. This enables gas turbines to operate efficiently under extreme conditions while maintaining high resistance to fatigue and fracture.

Nanostructured materials:

Nanostructured materials, such as metal nanowires, show how understanding dislocations at the nanoscale improves material properties. Dislocations in these materials can move more easily, offering better ductility and increased strength. This understanding is crucial for developing materials with specific properties for advanced applications, such as nanoscale electronic devices.

5. Future prospects

Cottrell's work continues to inspire modern materials science research. Advances in numerical simulations and experimental techniques allow his theories to be tested and refined, exploring emerging materials and advanced technologies. Researchers strive to apply dislocation concepts to develop materials with improved properties, integrating recent discoveries to meet current and future technological challenges.

This chapter illustrates how Sir Alan Cottrell's contributions have shaped our understanding of dislocations and their impact on materials science, with practical applications that continue to revolutionize modern industry.

Conclusion

Summary of key points

This book has offered a comprehensive exploration of dislocations, starting with their historic discovery in the early 20th century, which revolutionized our understanding of material deformation mechanisms. We have defined dislocations, detailed their structures, and explained the different types, including wedge and helical dislocations. The mechanisms of dislocation displacement and their impact on the mechanical properties of materials, such as strength, ductility and fatigue, have been clearly outlined. Modern observational techniques, such as transmission electron microscopy and X-ray diffraction, were reviewed, as were modeling and simulation methods. Case studies illustrated how dislocation management is essential for optimizing materials in practical applications ranging from gas turbine superalloys to nanostructured materials.

Future prospects

Research into dislocations continues to evolve, with promising developments in the understanding of nanoscale phenomena and the ability to design materials with tailor-made properties. Recent innovations in atomistic modeling and experimental techniques are paving the way for more robust, high-performance materials, notably in the aeronautical, automotive and biomedical fields. Emerging trends include the integration of dislocations into the design of advanced materials, and the use of nanotechnologies to manipulate mechanical properties on an even finer scale. Future research should focus on improving the durability of materials and optimizing their performance by exploiting dislocation mechanisms.

Lexicon

- **Dislocation**: Linear defect in a crystal lattice, influencing the mechanical properties of materials.
- **Slip**: Movement of atomic planes through the crystal lattice under stress.
- **Plastic deformation**: Permanent change in the shape of a material under stress.
- **Slip plane**: Specific plane where dislocations move more easily.
- **Edge dislocation**: Dislocation created by inserting an additional plane of atoms.
- **Screw dislocation**: Dislocation where the lattice is spirally deformed around the axis of the dislocation.
- **Burgers vector**: Vector describing the local shift in the crystal lattice caused by a dislocation.
- **Slip direction**: Direction in which dislocations are most likely to move.
- **Transmission Electron Microscopy (TEM)**: Technique for observing the internal structures of samples.
- **X-ray diffraction**: a method for analyzing crystal structure.

- **Atomic Force Spectroscopy (AFM)**: Technique for measuring the forces between a probe and the sample surface.
- **Atomistic modeling**: Simulating atomic interactions to study materials.
- **Continuous models**: Approaches that treat dislocations as macroscopic entities.
- **Molecular dynamics**: A method for simulating atomic motion.
- **Simulation software**: Programs to model dislocation behavior.
- **Superalloy**: Metal alloy designed to withstand high temperatures and severe conditions.
- **Precipitate**: Solid phase dispersed in a metal matrix that reinforces the material.
- **Nanowires**: Metal wires of nanometric size.
- **Nanotubes**: Tubular carbon structures.
- **Ductility**: Capacity of a material to deform plastically before breaking.
- **Hardening**: Heat treatment consisting of heating a metal to a high temperature and then cooling it rapidly to increase its hardness.
- **Tempering**: Heat treatment after quenching to reduce brittleness and improve ductility.

References

1. Allen, M. P., & Tildesley, D. J. (1987). *Computer Simulation of Liquids*. Oxford University Press.
2. Bhushan, B. (2004). *Springer Handbook of Nanotechnology*. Springer.
3. Cottrell, A. H. (1953). *Dislocations and Plastic Flow in Crystals*. Oxford University Press.
4. Cottrell, A. H. (1964). *The Strength of Solids*. Butterworths.
5. Cottrell, A. H. (1970). *The Mechanical Properties of Materials*. Macmillan.
6. Cottrell, A. H., & Oakeshott, L. H. H. B. (1980). *The Theory of Crystal Dislocations*. Academic Press.
7. Cullity, B. D., & Stock, S. R. (2001). *Elements of X-ray Diffraction (3rd ed.)*. Addison-Wesley.
8. Duan, X., & Lieber, C. M. (2000). *Nanowires for electronics and optoelectronics*. Journal of Materials Chemistry, 10(7), 1675-1688.
9. Frank, F. C., & Bilby, B. A. (1950). *On the exchange of dislocations between crystals*. Proceedings of the Royal Society of London. Series A, Mathematical and Physical Sciences, 199(1055), 111-131.

10. Gleiter, H. (1989). *Nanostructured materials: Basic concepts and physical properties*. Proceedings of the Royal Society A, 426(1877), 227-242.

11. Hirth, J. P., & Lothe, J. (1982). *Theory of Dislocations (2nd ed.)*. Wiley.

12. Hull, D., & Bacon, D. J. (2011). *Introduction to Dislocations (5th ed.)*. Pergamon Press.

13. Kocks, U. F., & Mecking, H. (2003). *Physics and phenomenology of strain hardening: The FCC case*. Progress in Materials Science, 48(3), 171-273.

14. Liu, Y., & Zhang, Y. (2004). *Carbon nanotubes: Properties and applications*. Advanced Materials, 16(5), 286-294.

15. Murray, J. L. (2000). *High temperature superalloys: Properties and performance*. Journal of Materials Engineering and Performance, 9(4), 403-412.

16. Nye, J. F. (1953). *Some geometrical relations in dislocated crystals*. Acta Metallurgica, 1(2), 153-162.

17. Pitts, J. R., & Smith, B. C. (2005). *Simulation of Defects in Crystalline Materials*. Cambridge University Press.

18. Ritter, J. R. (2001). *Heat treatment of nickel-based superalloys.* Heat Treatment of Metals, 28(3), 142-153.

19. Sohn, Y. H., & Rogers, R. R. (1991). *The role of precipitates in strengthening nickel-based superalloys.* Journal of Materials Science, 26(7), 1941-1950.

20. Suresh, S. (1998). *Fatigue of materials.* Cambridge University Press.

21. Wang, Y., & Chen, M. (2007). *Mechanical properties of nanowires and nanorods: A review.* Materials Science and Engineering: R: Reports, 58(3-5), 118-143.

22. Xie, Y., & Liu, C. (2008). *Mechanical behavior of nanostructured materials.* International Journal of Plasticity, 24(4), 655-669.

MIX
Papier aus verantwortungsvollen Quellen
Paper from responsible sources
FSC® C105338

Printed by Books on Demand GmbH, Norderstedt / Germany